Robert Main

The British Navy in 1871

Elibron Classics
www.elibron.com

Elibron Classics series.

ISBN 0-543-70638-9 (paperback)
ISBN 0-543-70637-0 (hardcover)

This Elibron Classics Replica Edition is an unabridged facsimile of the edition published in 1871 by Smith, Elder and Co., London.

THE

BRITISH NAVY

IN 1871:

By ROBERT MAIN,

AUTHOR OF "DOCKYARD ECONOMY," "NAVAL FINANCE," ETC., ETC

LONDON:
SMITH, ELDER AND CO., 15, WATERLOO PLACE.
1871.

Price One Shilling.

PREFACE.

The uneasy state of public affairs makes an intimate knowledge of the condition of our naval forces a matter of grave importance, and the writer has aimed at affording authentic information on the subject.

The substance of the following pages appeared in *Colburn's United Service Magazine*; but the whole has been revised, and some additions have been made to describe the present condition of the British Navy up to the latest date.

R. M.

THE BRITISH NAVY

IN 1871.

THE difficulties of European politics all point in the direction of war. We may have hugged to ourselves for a while the comfortable delusion that our insular position secured us exemption from attack; but we have now to confront realities instead of probabilities, and are compelled to ask seriously, as a question of moment, "Are we ready to meet any eventuality?"

The temper of this country has been tried, and has not been found wanting, for all classes of Englishmen feel that, if it is necessary to go to war to defend our national honour, fight we will; and, although there is an impression abroad that we are solely engrossed with our commercial interests, and too peace-loving to draw the sword, there is a latent force and determination to resist dictation and avenge insult, which, when roused to action, would prove that the might of England, if it has slumbered, has lost none of its puissance, and would be put forth when needful with even more than its wonted vigour and persistence.

The Navy constitutes our first line of defence, and our strongest force in attack; it is, as it has always been, the right arm of England's power; and never in our country's glorious annals have we possessed a more powerful fleet, more complete and efficient armaments, or more gallant and better trained seamen than at the present time.

A retrospect of ten years will be the best way of understanding the present position of officers and seamen in the Navy.

Before the days of iron-clads, the force of men required for the Navy was much larger than it is now. While in 1860 there were as many as 250 ships in commission, there are not at present more than 200 distributed over foreign and home stations; but whatever has become of the vessels, the officers remain. It is, indeed, certain that, during the last ten years, the ranks of officers have not been much thinned, though the number of men has been gradually reduced. As vacancies have occurred in the ranks of officers, they have been filled up, and, as a result, it is found impossible to give them employment. The old days of large squadrons, headed by a three-decker with a thousand men, are over; but, unfortunately for the contentment of the Service, these fine old days have passed away too quickly. The Naval Service has undergone successive changes, during the last ten years, which have left hardly a feature unaltered, and have in each case materially affected the welfare of the British sailor. In the old days sailors were full of frolicsome fun, with a possibility of mischief; a ship would come home and land a cargo of men ready to be fleeced to their heart's content, and well worth fleecing. The crew of the "Hannibal," paid off late at night twelve or thirteen years ago, included men, each with fifty and sixty pounds in his pocket, who roamed the streets of Portsmouth the next morning without a penny left them. This, however, was not so attractive a sequel to the cruise of the old tar as getting into Devonport, and hiring all the cabs in the town for a short cruise round the place, all hands outside, and the inside reserved for any "lubbers" who liked to take them. Happily, those days when sailors were the victims of crimps are well nigh over, and with them the old seaman has vanished; the old hands of ten years ago are now out of the field; the stoke-hole has superseded the decks and rigging, and the artisan holds a higher rank than the seaman in the Navy. Thus, there has not been found much difficulty in reducing the number of seamen year by year during the last ten years to something like the limits actually required of service.

With regard to the reserve of seamen in the ports, Mr. Childers has given the country some rather startling information. He stated in very plain language, that he had "roused out" from every hole and corner of these home ports all the men who for years had continued to evade sea service. And one of the first results of this rousing out was to find that, in the "Agincourt" which he commanded in 1869, there were five petty officers, one of whom had been eighteen years in the Service without ever having been to sea, and another with fifteen years' service was in a similar position; they had spent the whole of their time in harbour. The order for immediate preparation for sea-service seems to have had such an alarming effect upon some of the men that four claimed their discharge at once, and it appeared, on looking into the case, that one had been nineteen years in the Service, fourteen consecutive years of which he was in harbour; that another had been sixteen years in the Navy without having ever been to sea; that the third had been twelve years, and the fourth more than seven years without ever having been to sea. Will it be believed that these men were all continuous service men, retained in the Service with special pay and privileges for the benefit of retaining their valuable assistance? I need not here venture to pursue the subject further; but there seems very little doubt that these men were only a sample of many for whom the country had for years been paying a fancy price, as if to ensure their not having any knowledge of a sailor's life or duties, and never doing more for the country than looking after rotten ships, whose proper and certain fate now is to fall under the auctioneer's hammer. It is scarcely possible to state calmly the fact of the country having for years paid large sums of money to keep useless men to look after rotten stores which were preserved for the repair of rotten ships. Then again, in the third part of the reserve, the Coast-guard, there were found anomalies requiring investigation. This branch of the Service is known to be of great value, and is regarded as a most important reserve of highly trained seamen whose conduct has obtained them a short respite from hard sea-work, and whose character as seamen has earned them consideration as a valuable corps for the service of their country. But when the real test of usefulness was applied by ordering this force to prepare for immediate sea-service, it was found that *no less than* 600 *men* were unfit for service at sea.

Amongst naval officers, the anomalies were more striking than in the case of sailors. So flagrant were they, indeed, that the naval "retirement" scheme of last year was imperatively called for. How it will work remains to be seen; but its terms, notwithstanding some hard cases, seem fair, and have been received with a considerable amount of favour by all classes of naval officers whom it affects. An enumeration of its principles will be sufficient to show clearly the working of this scheme.

The origin of a scheme of retirement for naval officers was not due to the present First Lord of the Admiralty, the necessity for it having been felt years before. The inconvenience of overcrowded lists, the grievance of slow promotion, and the inability to give appointments with sufficient regularity to ensure a proper amount of active service to every officer in the Navy had been felt for years. But these difficulties, though smoothed, had not been overcome; and their existence constituted a dangerous defect in the machinery of the Navy. In 1860 and in 1866, schemes for improving the retirement and promotion of officers had been carried out, but proved ineffectual. They had been devised with a view to confer a boon on the Navy, without prejudicing existing interests in the Service, and they had consequently proved abortive. Although retirement was facilitated, it was only in a very slight degree compulsory: it did not affect all classes alike, and in the end it failed to secure any permanent benefit. Another disadvantage was, that additions were made to the lowest ranks in the Navy at the usual rate every year; so that in a very short time the appointments exceeded the retirements, and the inevitable block reappeared. At last, a fatal disadvantage became apparent, which proved insuperable. The gradual increase in the number of iron-clads led to a corresponding diminution of wooden vessels, and opportunities for appointments in the Service became extremely limited. Not that iron-clads required fewer officers than the old wooden ships; on the contrary, they required more. But as the absolute necessity for more iron-clads arose, so the conviction gained ground that wooden men-of-war were useless. Wooden ships were consequently condemned wholesale; the number of ships kept in commission was seriously reduced; appointments became more and more rare; and utter stagnation was the inevitable result of this combination of unavoidable mischances. Thus, two years ago, the state of things in the Navy, as regarded promotion, was worse than ever, and last year Mr. Childers stated, as an unquestionable reason for introducing a new retirement scheme, that the maintenance of a large number of unemployed officers was open to the most grave objections. "In the first place," he said, "it is very uneconomical; in the next place it is very mischievous, by creating discontent among the officers who are constantly on half-pay, and by causing agitation for increased pay; and, thirdly, it produces great inefficiency, because, in these days, if an officer is on shore for a long time he gets behindhand with the improvements that are always going on."

This statement is, of course, unanswerable. The changes in gunnery and ship-building are so rapid and so novel as to make it impossible for an officer to perform his duties effectively unless he has a good knowledge of them. Let any person accustomed to the old style of wooden vessels, be he officer or landsman, take the trouble to go over a modern iron-clad like the "Monarch," and then let him fancy what it is to take command of such a vessel. He will find nearly everything altered; the very manning of the guns is totally different. He will find telegraphs, and all sorts of mechanical and scientific accessories, of whose names, possibly, he was altogether ignorant, in constant use; his education would have to commence afresh, and without taking great pains, he would find himself virtually superseded, for all practical purposes, by younger but more experienced men. What Mr. Childers felt, and what officers in the Navy felt, was, that under a system of long half-pay, officers *had not a fair chance.*

"What are the facts," Mr. Childers said, "in regard to the employment of officers? I comprise under the head of employments, everything that can be so called, even including the civil appointments under the Admiralty; and, really, some of these appointments *would not exist but for the purpose of giving employment.* I omit altogether the retired List, and take only the Active List." He then showed that there were nearly five times as many flag-officers unemployed as on full pay; but it will be clearer to tabulate the facts, though they look unmistakeably repulsive.

Thus there were	17	Flag-Officers	employed, and	78	unemployed.
,,	92	Captains	,,	199	,,
,,	162	Commanders	,,	239	,,
,,	513	Lieutenants	,,	265	,,
,,	252	Navigating-Officers	,,	123	,,
,,	163	Chief-Engineers	,,	71	,,
,,	118	Chaplains, &c.	,,	33	,,
,,	390	Medical Officers	,,	135	,,
,,	157	Paymasters	,,	149	,,

These figures are very instructive; they show that out of 3,156 officers, only 1,864 are employed, while 1,292 are without employment of any kind. Bad, however, as this anomaly was, it was linked with worse; for the systems of employment, of half-pay, and of retirement, were so dissimilar that hardly two classes of officers appeared to have retired or have been put on half-pay upon the same system. As the First Lord very justly said, "The whole thing is a combination of happy accidents with individual fancies of particular Boards of Admiralty." So, like a knight of old, Mr. Childers set himself to redress this grievance; and, like many knights of old, he found he had not only an almost impossible task, but that in the doing of it he had to encounter difficulties of a very painful kind, and received more knocks than thanks for his exertions. What with the indignation of irate half-pay admirals and captains, and the impatience of young officers waiting for promotion, it is needless to say that their manifestations were not of a cheering character. Nor need this be wondered at, when we see how uncompromising the terms of retirement were. It was proposed to establish "an universal compulsory retirement for all ranks, with, in many instances, the option of an easy retirement." In fact, it was resolved that admirals of the fleet should retire at the age of 70; admirals and vice-admirals at 65; rear-admirals at 60; captains at 55; commanders at 50; and lieutenants at 45. And it was further proposed that officers who had not served for a certain period should retire—a flag-officer after ten years' absence from service; a captain after seven years; and a lieutenant after five.

No such comprehensive scheme as this has been within our recollection introduced into the Naval Service. That amongst many officers it spread consternation may well be imagined, but by the majority it could only be regarded as a boon. By its provisions many officers, tired of a service in which they had taken so slight a part and in which promotion was so exceptional, were glad to retire from it on the equitable terms which were offered. In fact, the scheme was statesmanlike in its conception, being alike beneficial for the country, for the Service, and for most of the officers whom it affected.

This scheme has of course materially affected the number of men serving in the Navy. The whole reduction carried out last year numbered 2,000, of which 500 were officers, brought under the terms of the retirement scheme, the remainder consisting of 500 seamen, 700 servants and artisans, and 300 boys. To show that this slight reduction was justifiable, it is enough to remark that the reserve in our ports now available, in addition to the 32,850 men actually serving afloat, is about 28,000 men. About a quarter of this number is in the home-ports and the coast-guard, employed in the harbour and other ships; the number of Marines on shore is about 6,200, and the rest consist of supernumeraries and coast-guard on shore, numbering over 12,000 men. Thus, it is clear that, for all the ships which would require to be commissioned in an ordinary way, there is a sufficient supply of men. Indeed, for the Navy in its normal state, a force of 60,000 is sufficient for all practical purposes. But a difficulty, which had been foreseen, is just now felt seriously: it is this, the state of the reserves, for service in the event of an emergency, is not so satisfactory as is desirable. We have the Royal Naval Reserve, and the Royal Naval Coast Volunteers; but admirable as is the spirit which animates both these bodies, it is felt that reserves, to be effective, must be organized on a broader basis and on a more extensive scale. To show that this difficulty had been appreciated, Mr. Childers proposed last year the establishment of a Seaman Pensioner Reserve; this met one part of the difficulty, by offering inducements to men retired at an early age from the Service, to keep up their connection with it, and retain in some degree their professional knowledge as efficient gunners and sailors, in the event of their being wanted in time of war. But this scheme will only partially meet the difficulty which, at the present time, presses with especial force, and will, no doubt, receive special attention. What is evidently wanted is a more elastic system than the present, which shall provide for an easier flow of men into and out of the Service; enabling them to retain for some time after leaving the Navy that knowledge they have acquired of their duties, and especially providing for a more complete and harmonious union between the Merchant Service and the Royal Navy.

But men are not of much use without ships, and any slight alarm which may now be felt by the country is, to some extent, excusable; for, not only in 1869, but last year, ships both large and small were sold off without scruple when they were reported to be useless, and our fleet at present has been reduced to a very compact size. The maintenance of a flying squadron was found so advantageous, that not only were certain reductions carried out at some of the stations, but, had it not been for the war, another squadron would have started last autumn.

Our Navy now consists of nearly 600 vessels, including every class from a cruiser or gunboat to a man-of-war; and there are in active service about 200, now in commission, of which 25 are iron-clads. The most important squadrons in the Mediterranean, the Channel, and the Coast-Guard consist at present of the finest vessels in the Navy. It is interesting to trace the progress made in replacing obsolete and useless vessels by those of the new construction; but in reviewing the efforts of the past year to provide new vessels for the Navy, two painful circumstances at once confront us,—the loss of the "Captain," and the resignation of Mr. Reed. It is not within the scope of this pamphlet to enter into either of these controversies; they can only be noticed as disasters, marring an otherwise energetic and prosperous attempt to reinforce the Navy with efficient vessels. One special point, however, requires attention here, as it is of a very important character.

Shipbuilding in our dockyards had, for many years, been carried on in a fitful, irregular manner; each administration seemed to think itself justified in getting on as well as it could with the ships it found in existence, and when a panic or other causes arose, going in for shipbuilding with violent energy. This fitful policy—arising partly from the rapid and great improvements necessitated by the increased power of the armaments of our war-ships—has been, as may well be imagined, the very reverse of what is either economical or judicious. It has led to ships being built in a hurry which were not wanted; to the admission of an excessive number of artisans at one time, and their necessary wholesale dismissal at another; and to the Navy being generally in want of ships of the right sort. As evidences of such defects, we have only to refer to the fact of Sir John Pakington finding himself compelled to build, all of a sudden, fifteen frigates, which became obsolete almost before they were completed; and to the dockyard dismissals of the last three years. Immoderate work, immoderate stagnation of work, and immoderate employment or dismissal of artisans, have been the fruit of this shuttlecock sort of policy. This state of things Mr. Childers has already attempted to set right, by laying down what he considers the minimum annual rate at which shipbuilding should be maintained, and by following it out.

After mature consideration, he arrived at the conclusion that to keep the Navy up to its due strength in ships, so as to maintain a minimum standard of efficiency, it would be necessary to build about 20,000 tons of shipping annually. This steady increase to the strength of the Navy would suffice to supply deficiencies caused by the annual waste or decay of ships, and would obviate the necessity, except in a time of emergency, of resorting to any unusual means for increasing their number. This annual increase he proposed should be divided between the armoured and unarmoured vessels, by building 12,000 tons of the former, and 8,000 tons of the latter, class; and, with the view of making use of the dockyards and dockyard labour as much as possible, he proposed that about a fifth of the work only should be done by contract. To show how important an addition this annual increase would be to our fleets, I need only state that it represents three new iron-clads, a frigate, a corvette, and six smaller vessels annually. This would, indeed, be a very satisfactory plan if properly carried out; its cost would be roughly about two millions and a half: in other words, about a quarter of the sum annually voted for the Navy would be devoted to the solid increase of its power.

But to understand a little more clearly the advantages of this policy, it is necessary to know, practically, how our Navy stands in regard to ships, especially iron-clads, at the present moment. This time last year there were forty-seven armoured vessels of all descriptions, of which thirty-seven were complete, and ten were in course of construction. An enumeration of their names will be all that is necessary, to show in what position we now stand.* The ten new vessels under construction are the "Iron

* See list at p. 31.

Duke," "Sultan," and "Glatton," at the Royal dockyards; the "Swiftsure," "Vanguard," "Triumph," and "Hotspur," which are being built by contract; and lastly, the three new ships laid down by Mr. Childers, the "Devastation," "Rupert," and "Thunderer." This was the condition of our iron-clad fleet a year ago. To show how far the Government has endeavoured to give a practical effect to its shipbuilding policy, and what progress has been made during the past year, we can state that of these ten unfinished vessels five were launched, the "Vanguard," "Iron Duke," "Hotspur," "Sultan," and "Swiftsure," and, as will be seen by the following statement, five new ships were laid down; namely, the "Fury," "Cyclops," "Gorgon," "Hecate," and "Hydra," all turret-ships; of which two—the "Fury" and "Hecate"—are to be built at the Royal dockyards.

We start, therefore, in the present year with fifty-one iron-clads, of which ten are not quite complete; and but for the loss of the "Captain," we should have had fifty-two. The finest vessel added to the Navy last year is the "Sultan," which is not unlike the "Hercules," but is superior to it from being able to carry its armament in a way that makes its guns more powerful. The outbreak of war has caused the Government to push forward ship-building with great energy during the past few months, and it is by no means improbable that as many iron-clads will be added to the Navy this year as were added last year. Great progress has been made with the "Devastation," "Glatton," and "Triumph," and we may expect to hear of their being launched before many months are over.

Before leaving the subject of ships, I ought to notice the launch of the "Plucky," the "Comet," and

"Blazer," little gun-boats built like the "Staunch," carrying in their bows one gun, mounted and fired after the fashion of a Moncrieff battery. These gun-boats are a great addition to our naval strength: the loss of the gun-boat "Slaney" last year was a misfortune to the Service and a loss to the country. It is proposed to build two armoured gun-vessels at Chatham, the "Scourge" and "Snake," designed probably with the view of acting like the gun-boats on the Seine, near Paris, which did good service in the sortie on the 30th of November; lastly, there are some wooden vessels of various sizes now under construction. The "Blonde" is an iron screw-vessel being built at Portsmouth; the "Raleigh" has just been laid down at Chatham, and is an iron screw-frigate cased with wood; the "Robust" is an 81 gun screw-frigate, on the stocks at Devonport; the "Bulwark" is an 81 gun screw-frigate laid down at Chatham a long time ago, which last year it was proposed to convert into an iron-clad, but upon investigation it turned out that the experiment would be very costly and not by any means satisfactory, and the idea has been therefore abandoned. Four small vessels—the "Lively," "Osborne," "Vigilant," and "Woodlark"—conclude the ship-building work which the Government has now in hand.

With respect to our naval armaments, it will not be forgotten that the accident in the "Hercules," added to other untoward circumstances, induced Mr. Childers to look closely into the present state of the heavy ordnance for the fleet. Last year he expressed himself satisfied with all guns of a 12-ton 9-inch calibre and under; but suggested that the heavier ordnance for the Navy was not satisfactory. The 18-ton gun was not quite trustworthy; but, on the other hand, the 25-ton gun had been proved thoroughly efficient. In other words, while the ordnance of the most powerful and least powerful iron-clads was good, the armaments of the "Hercules" and "Sultan" were not altogether satisfactory. Then, again, good as the 25-ton gun was for the "Monarch," "Hotspur," and "Triumph," it is not powerful enough for the new ships now being constructed, such as the "Devastation," "Fury," and others. For these vessels it has been determined to have at least 30-ton guns, which have yet to be constructed: it is not intended to open competitions for guns of this class. The chilled projectiles for the Navy are not quite satisfactory, several having burst; competition, therefore, is to be allowed for the best projectile. Captain Palliser, in discussing this question, thought that the failures were not serious enough or sufficient in number to warrant a re-opening of the controversy of guns and projectiles; but we cannot be too certain on such vital points in warfare as armament and ammunition, though to enter into a course of experiments heedlessly or unnecessarily, when every round from a 600-pounder costs the country about nine pounds sterling, would be wasteful. It is satisfactory to know, on good authority, that our service-guns are safe. Cases of bursting are very few, and there are guns now in use which are in no way damaged after having fired 1,000 or 2,000 rounds. But with the knowledge which recent experience has given that the best gun is one of the safest steps to victory in warfare, we cannot afford to lose any opportunity of placing our naval ordnance in the highest state of efficiency. A step in this direction was taken last year by the issue of an order from the War Office for the construction of a 35-ton gun, which was completed a few weeks ago, and is to be simply an enlargement of the service gun; that is, it consists of a series of coils. We can form some idea of the gigantic size of this gun when we know that the outer coil consists of a bar of iron nearly equal in length to the height of the Monument. This enormous piece of ordnance weighs 35 tons 7 cwt., and will throw a 700lb. shot, which will pierce iron armour 15 inches in thickness: the ordinary charge of powder is calculated at 120lb. The tube of the gun, which is about 16 feet long, was brought out of the factory in which it has been rifled and fixed, muzzle downwards, in a pit, under a powerful crane; the heavy breech-piece—a mass of iron weighing 15 tons—was heated nearly to redness in order to expand the metal, and the heated mass was lifted and dropped like a cap over the breech of the perpendicular tube. The moderate calibre of the new gun is designed for penetrating, and the shot will probably be three times the length of its diameter. This monster cannon now only requires to be vented and proved to be ready for service, which it is confidently expected to be immediately. The estimated cost is 2,500*l*. Sir William Armstrong's estimate for a 35-ton gun was 3,500*l*., and Sir Joseph Whitworth's 6,000*l*.

One very important point was taken up last year in connection with naval ordnance, namely, the best way of protecting ships' decks, which are very vulnerable when exposed to danger from a plunging fire. Several experiments were conducted at Shoeburyness, which proved on the whole satisfactory to the Navy, by showing that the penetrability of such decks as those of the "Monarch" and the larger iron-clads was not so great as was imagined. Some interesting experiments took place at Liège in the course of the summer, the principal object of which was apparently to ascertain how to increase the durability of small guns, especially those of bronze. MM. Montefiori and Kunzel had invented some new process which was considered efficient; their idea being that by the mixture and addition of phosphorus, old bronze guns might be strengthened, and that with guns, as with men, phosphorus might be found to be the *vis vitæ*—the secret of lasting power.

A very successful and interesting series of experiments was conducted during the past year at Shoeburyness, in order to test the strength of Palliser shells with the heavy naval ordnance, and also the bronze field-gun for India. But the experiments which excited most interest were those with the mitrailleuse; since then the trials of real warfare in the last few months have determined the value of this arm. It is satisfactory to know, in regard to experiments, that the old-fashioned way of testing iron plates has been superseded; instead of using the old smooth-bore gun to test the plates of the "Devastation" and "Glatton" last year, the 7-inch rifled gun was used with chilled shot.

We ought not to omit all notice of the torpedo experiments which have lately been conducted. Some reproach has been cast on the Government, not without reason, for having given too little attention to torpedoes; the subject has now, however, been taken up with vigour, and, after the melancholy accidents at Cuxhaven, will not be abandoned until our shores are protected against a hostile fleet by these formidable submarine engines of defence.

As regards the dockyards, the rapid progress made in the extension works at Portsmouth and Chatham require notice. When these works are complete, they will afford the most ample accommodation for our largest ships; but it will be some years hence. By means of an excellent plan devised by Colonel Clarke, C.B., the Director of Works, the Marines have been employed on these works with the most successful results. At the commencement of the year, the reductions which so annoyed the Government and the Dockyards, had been effected so as to bring down the establishments to their proper state of efficiency.

To complete this brief survey of our naval power, it is necessary to describe the administration employed in its maintenance. The disastrous failure of the French Intendance in the present terrible war should be a lesson to the Government of this country, showing the vital importance to the welfare and power of the nation of an efficient and vigorous Administration. Hitherto, that of the Board of Admiralty has had the worst reputation; it is not surprising, therefore, that one of the first acts of the present Government was to reform the Admiralty. The Board having been attacked violently, Mr. Childers took upon himself the responsibility of guiding naval matters, and assumed the office of a Minister of Marine. The Secretary of the Admiralty became, as a next step, an active assistant to the First Lord, and, in effect, a sort of deputy to him. To him has been assigned the control of all financial questions; to him are entrusted the arrangements for making purchases and for keeping the accounts; and he has to deal with all matters affecting naval finance. In short, as Financial Secretary, he has to keep a sharp look-out after economy in the naval expenditure; in this he is assisted by two other members, whose duties are well-defined, but who are subordinate to him. As the Financial Secretary now acts as a sort of Treasurer to the Navy, the Controller, who had before been a subordinate officer, was made a member of the Board, and now takes the management of the dockyards, and directs the construction and repair of ships. A naval officer still looks after the interests of officers and seamen, determines questions of promotion and pay, and manages the *personnel* of the Navy. Thus each member has definitely assigned duties, but is subordinate to the Minister, to whom all questions of policy and matters involving a departure from established rules have to be referred; so have endless discussions been attempted to be replaced by rapid and definite decisions.

This change, so long insisted upon, was acquiesced in at once without a murmur. In an economical point of view there cannot be much difference of expense between the two Boards. But reform did not stop here. No sooner had the Board been put on a proper footing of subordination to the First Lord, than immediate steps were taken to re-organize the departments of the Admiralty. The familiar proverb, "give a dog a bad name and it will stick to him," may well be applied to this branch of the Civil Service. If inefficiency was known to exist in the Civil Service, the Admiralty was sure to be singled out as the scapegoat for the sins of the other branches. The Admiralty was named as the great exemplar of inefficiency and of indifference to public interests. But there were good grounds for this; and there was sufficient justification for the prompt and apparently severe measures adopted by Mr. Childers. Every blow aimed at the Admiralty can be traced to the disasters of the Crimean war; and it must be admitted that mistakes both numerous and inexcusable were then committed. There was a complete absence of responsibility. Men were found holding both important and subordinate positions, who were thoroughly incapable, and alike impervious to a sense of the importance of their duties and to criticism.

The disastrous defects of the administrative system fifteen years ago, led to the adoption by the Government of a new principle of entry into all public appointments. A Commission was appointed, and tests were established to be applied to every future candidate for the Civil Service. It was expected that a better class of men would be obtained by this means, and that the Service would in a very short time attain a higher tone and character. But what is the result? In the Admiralty it has been necessary apparently to re-construct, from top to bottom, the machinery for carrying on public business. Fifteen years ought to have sufficed to the Civil Service Commission to do its work among the subordinate branches, if it had not been able to affect the Board of Admiralty. But no sooner was the Board placed on a new footing, than Mr. Childers found he must proceed to re-cast the subordinate departments, and introduce alterations in order to produce the very results anticipated from the work of the Commission. It may be asked, was the Commission to blame, or had the Admiralty contrived to defeat the Commissioners? The answer to this question is very simple. By the establishment of a more rigorous test of capacity it was hoped to ensure the entry of efficient men into the service; but fifteen years have proved to be too short a time to allow their influence to be felt. It is hardly too much to say that not more than one man in ten, who had entered the Admiralty with certificates from the Civil Service Commissioners, is now in a position involving the slightest amount of responsibility. There are, it is true, many men who passed the Civil Service examinations when first established, but who have not been promoted, and have not been passed over, but who have no chance of promotion for years to come. The natural consequence of such a state of things is at once apparent. All the posts involving responsibility—or at least, a majority of them—are still filled by men who were never subjected to any test worth mentioning. The Civil Service Commission has done little as yet for the Admiralty but introduce well-trained men into the service, who have, with but few exceptions, been employed on mechanical work which it is now admitted might well be done by coarser hands. To say that the Commission has failed in supplying the Admiralty with efficient clerks would be untrue, for the improvement has never been tried fairly; and in considering the reforms commenced by Mr. Childers, this fact must be borne in mind. The same class of men which held all the high positions during the Crimean war retained them last year; the same traditions prevailed; and, with but few exceptions, the same obsolete habits not only existed, but were even cultivated. Work was done in the old cumbrous way; the same ponderous and inelastic machinery was used, and the old routine, so disheartening to the public which had to do business with the Admiralty, and heartbreaking to those in its service who wished to do their work promptly and efficiently, still held its sway. To effect what the Civil Service Commission had failed in; to sweep away the official cobwebs which obscured public business; to do, in fact, in a rougher and more thorough manner what had been attempted by gentler methods, was the task to which Mr. Childers set himself.

The question will naturally be asked, how has he succeeded? But partially will be the answer of most critics; and in many respects it will, perhaps, be added unevenly. It cannot but be admitted that so far as reform in the departments under his charge has gone, the results achieved by Mr. Childers have been as yet only partial. Nevertheless, he has introduced a reform in the administration of Admiralty business, so real and so powerful, that its defects, patent as they are, are not so much material blots in his scheme as accidents which the new system must, in the natural course of time, inevitably remove. The reform which he has commenced is based on two principles—economy of work, and individual responsibility. In the management of such a vast and complicated establishment as the Navy, it is of the utmost importance that the various branches of its work should co-operate with one another naturally and without interference; that each branch should be self-contained and manage its own work in an efficient manner; and that each person in the Service should be responsible for his own work. Starting with these leading principles, and with the knowledge that they had been neglected in most departments of the Admiralty, Mr. Childers in proceeding to reform the Admiralty administration applied the two principles of economizing the work and imposing individual responsibility. Dealing with them in their order, let us see what was found when first the attempt was made to economize the work and separate it into its proper and natural divisions? The first discovery was that one of the most common and flagrant breaches in the economy of work was the constant repetition of the same work in different branches and departments. Departments checked each other; branches checked each other; and rooms even checked each other. In some cases it appeared this was done purposely; but in others unintentionally. Two departments were discovered to be doing the very same kind of work, without any practical effect, and without each other's knowledge. This was easily accounted for. Both these departments were in the Admiralty it is true; but the one had little to do with the other, and they were not under the same roof. Had the work been done twice over, with the distinct purpose of providing a check, there would have been some excuse; but in this case no such attempt was made, for neither department knew what the other was about.

One of the first things Mr. Childers did, was to reduce the staff of one of these offices. So astonished were the members of the late Board at this, that they asked, on one occasion, how it happened that in a department which they had reluctantly been compelled to increase, the present Board was able, without any difficulty, to make its first reductions. The reply which Mr. Childers gave was convincing. "There was indeed a great necessity for writers' work at that office; but a great deal of the work was useless." In fact it had been found *that by economizing the work it was possible to reduce the staff*, and that the efficiency of the work done was not in the least injured by the application of a little wholesome economy. After all, the whole secret of this and other anomalies which were found to underlay the traditional system of Admiralty work, can be found in the baneful practice of isolation. Isolation of men and work in Government offices is the secret of any want of success in their work, and of the ease with which mistakes are found by men, like Mr. Childers, who are sufficiently determined and energetic to remove them. Isolation has indeed been the bane of the Admiralty departments. In the case already referred to, this is palpable. Here were two branches of the same office, within a shilling cab-fare of each other, and both, without knowing it, engaged on the same kind of work. Each of these branches was as completely isolated as though it belonged to some office in no way connected with the Admiralty, and each kept to its own traditions, and its own system of doing work, as though, instead of being only the part of a machine, it were the whole machine itself. But it is possible to carry this complaint further: indeed, it is barely an exaggeration to say that the same state of isolation existed between rooms under the same roof. The feeble and effeminate habit of writing little notes from one room to another can only be excused or explained by supposing that there may be, in public offices, some notion that tenacity to a seat during office-hours is an official duty. Can it be wondered at, that under such a state of things, the business of the Admiralty in its inner life should have been obscured in a network of half intelligible correspondence, and that, in its intercourse with the outer world, this same network of correspondence should involve the most ordinary transaction; that personal communication should be avoided, and the practice of putting every trifle in "black and white" almost deified, and that, as a consequence, the most contradictory orders should be constantly given? To remedy these evils it was determined to bring the various branches of the office as closely together as possible under the eye of the Board, with the view of preventing the same work being done by two branches. But another evil had to be remedied, namely, the irresponsible position of the worker. For the highest, as for the lowest, class of work nobody seemed to be responsible. To such a height had this evil arisen, that every paper dealt with by a junior clerk was so covered with initials before it reached its last stage, that it was difficult to say who could be held responsible for any error which might arise upon it. More than half the duties of each office consisted, indeed, in simply repeating processes which had already been done. To remedy this defect, it was determined to abolish a whole class of senior men whose duties had consisted of little else than checking work which had already been checked. The class below them was entrusted with the responsibilities of its own work, and received a slightly increased scale of pay; but the junior class was not materially affected by this alteration, beyond having a slight increase in responsibility for its work.

To revert to the important subject of the Navy. With respect to the number and strength of our war-ships, it is clear that a force of fifty iron-clads is not sufficiently strong. Mr. Cobden, the most peace-loving of economists, laid it down as an axiom that the British Fleet should be as strong as the navies of any two European Powers. But, without endorsing his dictum, there can be no doubt that we cannot discontinue ship-building, and that it is necessary to keep up our dock-yards both for building and for repairs of ships, and also to maintain a sufficient force of workmen to meet all the wants of the Service in this respect. Then, as regards officers, it is essential that we should have a staff sufficient not merely for current requirements but for emergencies; while as regards seamen, it is no less necessary that the reserves should be ample and the men efficient. The Government has certainly not departed from its principles of efficiency, nor, in straining after economy, has it sacrificed any of them. Ship-building is not discontinued; for, as has been already shown, Mr. Childers, acting upon professional advice, has adopted a certain rate which he is determined to keep up year by year; and, to provide for the repairs of ships, as well as for building, an ample supply of men is provided.

The most startling reductions have been made in the ranks of officers; but, upon investigation, it will be found that these reductions will increase, rather than lessen efficiency. And with regard to the force of seamen necessary for the Service, it is satisfactory to know that it is both efficient and sufficient.

In fact, so far as reforms have yet been attempted, it is impossible to doubt that their tendency is to benefit rather than injure the Service; and to increase rather than lessen its efficiency. There have been errors; but how can they be altogether avoided in the execution of any reform? Such as have occurred, are rather in matters of detail than principle, and no attempt has been made, or is even necessary, to conceal them or evade their admission. In the Navy, the country wants a service complete and strong in all its parts; which shall be capable of affording protection to our commerce and our interests abroad, and of guarding our own coasts. The old idea, that mere numbers mean power; that large fleets, thousands of men, and acres of stores mean efficiency, has been proved to be erroneous. The Government, in boldly attacking this prejudice, has cleared the Navy of what did not represent real strength, cut away masses of useless and cumbrous top-hamper, and at the same time introduced an economy which has already materially lessened the burden of taxation.

The following list of our Fleet of Iron-clads must carry the conviction to all unprejudiced and impartial minds, that we are able to cope with any naval force that can be brought against us:—

Broadside Iron-clads (36 *in number*).

I. Hercules, Sultan.
II. Audacious, Invincible, Vanguard, Iron Duke, Swiftsure, Triumph.
III. Bellerophon, Lord Clyde, Lord Warden, Minotaur, Agincourt, Northumberland, Royal Alfred, Penelope, Repulse.
IV. Achilles, Royal Oak, Prince Consort, Caledonia, Ocean, Valiant, Hector, Zealous.
V. Warrior, Black Prince, Defence, Resistance.
VI. Pallas, Favorite.
VII. Enterprise, Research, Viper, Vixen, Waterwitch.
Added in 1870: None.

Turret Iron-clads (15 *in number*).

I. Devastation, Thunderer.
II. Monarch.
III. Glatton.
IV. Hotspur, Rupert.
V. Royal Sovereign, Prince Albert.
VI. Wyvern, Scorpion. Added in 1870.
At the dockyards: Fury, Hecate.
By contract: Cyclops, Gorgon, Hydra.

Not only in ships, however, but in officers, seamen, and guns, is the Navy in a better state, at the opening of the year 1871, than it has been for years.

THE END.

LONDON:
PRINTED BY SMITH, ELDER AND CO.,
OLD BAILEY, E.C.

Printed in Dunstable, United Kingdom

91219948R00030